Radix-4

Supriya Saste
Anil Sawant

Radix-4 Algorithm based Computations

LAP LAMBERT Academic Publishing

Imprint

Any brand names and product names mentioned in this book are subject to trademark, brand or patent protection and are trademarks or registered trademarks of their respective holders. The use of brand names, product names, common names, trade names, product descriptions etc. even without a particular marking in this work is in no way to be construed to mean that such names may be regarded as unrestricted in respect of trademark and brand protection legislation and could thus be used by anyone.

Cover image: www.ingimage.com

Publisher:
LAP LAMBERT Academic Publishing
is a trademark of
International Book Market Service Ltd., member of OmniScriptum Publishing Group
17 Meldrum Street, Beau Bassin 71504, Mauritius

Printed at: see last page
ISBN: 978-620-2-52409-4

CHAPTER 1
INTRODUCTION

1.1 Background

Arithmetic blocks play a crucial role in the digital and mixed-signal systems. The bottleneck of digital signal processing and many other digital systems is arithmetic blocks where addition and multiplication operations are the core units. As the VLSI technology reduces to nanometer size devices, power dissipation reduces and device speeds increase every year. However, silicon technology is reaching its physical limits, where the design sizes already have reached the atomic levels. In this thesis, three arithmetic operations viz., multiplication, addition and subtraction are implemented based on radix-4 algorithm.

Multiplication is a mathematical operation in which an integer is added a specified number of times. Multipliers are indispensable in modern electronic systems where high speed calculations are required such as FIR filters, digital signal processors and microprocessors etc. Multiplication based operations such as multiply and accumulate (MAC) and inner product are among some of the frequently used computations. Currently, time taken by multiplication operation is still the dominant factor in determining the instruction cycle time of a DSP chip and the demand for high speed processing has been increasing because of expanding computing and signal processing applications. Also the multiplier is generally the slowest element in the system.

Addition of numbers is the most basic operation of all arithmetic systems, such as subtraction, multiplication and division. In the basic addition structure, the addition time is proportional to the bit size of the numbers to be added since the next output digit depends on previous carry-out of each digit. This causes an unacceptable delay for many systems especially when the word-length of the input operands is high. Therefore, carry signal propagation must be eliminated in the arithmetic circuits. There have been various techniques developed for breaking the long carry chains of the arithmetic circuits. Basically, these techniques are based on redundant number representations.

Redundant representation means that a number is represented in more than one way in the number system. As an example, if the number base is radix-4 (each weight of the digit is a multiple of 4^n where n represents the nth digit) and digit set is selected to be $\{-3, -2, -1, 0, 1, 2, 3\}$. The redundancy in the number system can provide the carry chains in the addition algorithms to be broken. This makes the arithmetic operations independent on the bit size of the operands. The details of redundant systems and carry-free addition circuits are explained in details in the next chapter.

Redundant systems are not necessarily binary, but the numbers computed can be encoded as binary. As a result, more than one digit is required for the representation of 2 redundant systems when the system is implemented in binary logic. Hence, the redundancy in numbers means that the number structure is non-binary.

Considering redundant number systems in the number theoretic approach, carry propagation free algorithm development is achieved by using signed digit numbers or using carry save arithmetic. A signed digit system can be represented as A − B, where A represents positive digits and B represent negative digits. On the other hand, carry save format numbers can be represented as the addition of two numbers; $Z = \{S, C\}$, i. e., sum digits (S) and carry digits (C), respectively. Shortly, carry-save mode addition techniques may also be considered as redundant number addition techniques where negative numbers do not exist in the digit set. The details of redundant systems are analyzed in further chapter.

1.2 Importance of Topic Selected

With rapid increase in scale of integration, many sophisticated signal processing as well as video processing operations are being implemented on VLSI chips. Various arithmetic modules are used in such systems. The performance in very-large-scale-integrated (VLSI) systems such as digital signal processing (DSP) chips is predominantly determined by the speed of arithmetic modules like adders and multipliers.

This project deals with design and implementation of three basic arithmetic operations viz., multiplication, addition and subtraction based on Radix-4 algorithm. These arithmetic operations based on Radix-4 algorithm reduces number of iterations and helps to speed up operation. With increase in the speed of system, overall performance of system increases too.

1.3 Problem Statement

In order to perform fast operations, fast computational blocks are needed which will consume less power. So, the main problem is to design such computational blocks for arithmetic operations and to simulate it on suitable software and validate by hardware.

1.4 Objectives of the Project

- The main objective of this project is to design a scheme for fast computation based on Radix 4 algorithm which will be hardware –resource efficient using Verilog HDL.
- The project covers three basic arithmetic operations:
  - ➤ Multiplication
  - ➤ Addition
  - ➤ Subtraction
- Final implementation of these operations on FPGA platform. The test results of FPGA implementations validate the feasibility and effectiveness of scheme for designing Radix 4 based operations.

1.5 Thesis Outline

The work presented in this thesis deals with implementation of three basic arithmetic operations viz., multiplication, addition and subtraction based on Radix-4 algorithm and their simulation on FPGA platform. All three arithmetic operations are simulated in XILINX by using Radix-4 scheme. Radix-4 Booth algorithm is used to implement multiplication operation and addition, subtraction operations are implemented by using quaternary signed digit number system for which radix is 4.

This thesis is organized as follows: Chapter 1 presents basic introduction of topic selected. Chapter 2 presents the literature survey. Chapter 3 focuses on overview of redundant radix-4 system and how RR-4 number system is used. Chapter 4 provides detailed description of multiplication, addition and subtraction operation. This chapter gives clear idea about basics of multiplication, how radix-4 booth algorithm is used to implement multiplication operation with less number of partial products. Basics of addition/subtraction and how carry free addition and borrow free subtraction can be implemented by using RR-4 number system are presented in this chapter. Chapter 5 presents information of field programmable gate array. Corresponding simulation results are given in Chapter 6 followed by conclusion in Chapter 7.

CHAPTER 2
LITERATURE REVIEW

The speed of multiplication operation is of great importance in digital signal processing as well as in the general purpose processors, since the multipliers the key components of many high performance systems [1]. Radix-4 Booth's algorithm is an alternate solution to basic binary multiplication, which can help in reducing partial products by the factor of 2 [2].

Earlier ALU's adders were used to perform the multiplication operation. Use of multiplication operation in digital computing and digital electronics is very intense in the field of multimedia and digital signal processing application. Such applications demand great computation capacity [3]. Radix-4 booth multiplier has reduced power consumption than the conventional radix 2 booth multiplier [3].

Multiplication is more complicated than addition, being implemented by shifting as well as addition [4]. If more partial products are involved in multiplication operation, more time and circuit area are required to compute, allocate and sum the partial products. Recoding scheme used by radix 4 booth algorithm reduces partial products by half and helps to increase the speed of multiplication. Various types of adders that can be used for multiplication operation are listed in [4]:

1. Ripple Adder
2. Carry Look Ahead Adder
3. Carry Select Adder
4. Hybrid Adder

The good approach to implement multiplier is hybrid architecture of radix-4/-8, because radix-8 has low power consumption and partial products generated in this mode are less (N/3) compared to partial products generated in case of radix-4 (N/2). While computing partial products it is very difficult to detect 3B term and hence it becomes difficult to implement it on FPGA board. By comparing performances of two, [5] suggests to go for radix-4 multiplier. Radix-4 multiplier architecture is appropriate solution for portable multimedia applications [5] since it consumes low power and has high speed. The resource consumption of Booth radix-4 multiplier is 88.8% less than the Wallace Tree multiplier and the performance of Booth radix-4 multiplier is almost equal to the Wallace Tree multiplier [6].

Low power consumption and smaller area are some of the most important criteria for the fabrication of DSP systems and high performance systems. Optimizing the speed and area of the multiplier is a major design issue. However, area and speed are usually conflicting constraints so that improving speed results mostly in larger areas [7]. A serial radix-4 interleaved modular multiplier provides 50% reduction in the required clock cycles. In addition to the reduction in clock cycles, a parallel modular multiplier maintains a critical path delay comparable to the bit serial interleaved multipliers [8].

Multiplication operation can be considered as series of repeated additions. The repeated addition method suggested in past is slow that it is almost replaced by an algorithm which makes use of positional algorithm. The methods required to implement a high speed and high performance parallel complex number multiplier are implemented in [9], which make use of Radix-4 Modified Booth Algorithm along with Wallace tree. To enhance the speed of addition process carry save adder (CSA) is used in [9].

Addition is one of the basic arithmetic operations which is being used to compute all other algorithms like FIR, IIR. Many times adders are used in digital processors. Hence the speed of digital processor depends on speed of the adders used in the system [11]. With the binary system speed of arithmetic operations is limited by formation and propagation of carry. With binary system this design is easily possible when interconnection is less. With increasing number of inputs, interconnection becomes tedious work. The quaternary signed digit provides us to increase the logic levels. QSDN is the base 4 redundant number system [12]. The logic circuits based on this number system give less delay as compared to the binary circuits [12].

Quaternary Signed Digit number representations allow fast addition/subtraction which is capable of carry free addition and borrow free subtraction because the carry propagation chains are eliminated. Since the carry propagation chains are eliminated, it reduces the propagation time compared to radix-2 system [13].In quaternary signed digit number system, each digit can be can be represented by a number from -3 to+3. Using quaternary signed digit number system any integer can be represented in multiple ways [14].

CHAPTER 3

SCIENTIFIC BACKGROUND

3.1 Overview of RR-4 Number System:

Redundant arithmetic number systems are gaining popularity in computationally intensive environments particularly because of the carry-free addition/ subtraction properties they possess. This property has enabled arithmetic operations such as addition, multiplication, division, square root, etc., to be performed much faster than with conventional binary number systems. Redundant number systems are positional number systems similar to the conventional decimal system, but allow more values for a digit than the base.

This results in more than one representation for a number. Arithmetic operators designed using this number system achieve considerable speed improvement compared with operators designed using conventional number system, as the carry between the digits does not have to be consumed at every operation. Speed of an arithmetic operator is an issue that is directly related to the chosen architecture. Architectures and implementation styles are usually decided by the number system employed to represent numbers. Architectures are concerned with the direct hardware configuration of logic units (e.g., logic gates) for the physical realization of the arithmetic algorithm. Implementation style dictates the manner in which communication between different basic units of the architecture and the communication with the external world has to be achieved. For example, in conventional binary number systems, many implementation styles are possible such as bit-parallel (all bits at a time), bit-serial (one bit at a time), digit-serial (multiple bits at a time), word-parallel (multiple words at a time), etc.

In redundant arithmetic architectures two implementation styles are prominent. They are digit parallel and digit-serial. Digit-parallel is similar to bit-parallel (for binary number system) implementation. An interesting feature of this type of implementation is that both most-significant digit first and the least-significant digit first realizations can be easily achieved with low computational latencies.

3.2 RR-4 Number System:

In RR-4 number system the radix used is 4 and individual digits belong to the set, S= {-3, -2, -1, 0, 1, 2, 3}. There are more than one possible representations of the same integer in RR-4 number system. For example, [0 3 1] $_{RR-4}$, [1 -1 1] $_{RR-4}$, and [1 0 -3] $_{RR-4}$, all represent the number $(13)_{10}$.

This redundancy in number representation will be exploited to perform carry propagation – free addition, thereby allowing the parallel addition of four RR-4 numbers in $O(1)$ time, independent of the length of the numbers. Of the different possible representation of RR-4 digits, one possible way of writing the digits of set S using three binary bits for each digit are as follows, where the leftmost bit is 0(1) if the digit is positive(negative):

(-3)RR-4 = 101

(-2)RR-4 = 110

(-1)RR-4 = 111

(0)RR-4 = 000

(1)RR-4 = 001

(2)RR-4 = 010

(3)RR-4 = 011

The representation of any RR-4 number using binary bits can be visualized by a matrix of 0's and 1's as follows, where each column represents the respective RR-4 digit:

$(2 -1\ 0\ 3\ 1)_{RR-4}$ = 0 1 0 0 0

1 0 0 1 0

0 1 0 1 1

The topmost bit in each column indicates the sign of the digit, where 0 stands for positive and 1 stands for negative digit.

In this thesis we have implemented three basic arithmetic operations by using Radix– 4 algorithm. The general and detailed introduction of these operations is presented in further sections.

CHAPTER 4
DETAILED DESCRIPTION

4.1 MULTIPLICATION:

This chapter deals with the various algorithms of booth's recoding and the basic principles, architectures of adders that exist for binary addition.

4.1.1 Basics of Multiplication:

Multiplication is the most important and basic arithmetic operation carried out by ALUs. The speed of multiplication operation is of great importance in digital signal processing as well as in general purpose processors [1-3] [5]. While performance and area remain to be two major design goals, power consumption has become a critical concern in today's VLSI system design. Multiplication is a fundamental operation in most arithmetic computing systems.

Many digital signal processing methods use nonlinear functions such as discrete wavelet transform (DWT), discrete or cosine transforms (DCT). Because they are basically accomplished by repetitive application of addition and multiplication, these arithmetics determine the execution speed and performance of the entire calculation [2]. Digital multiplication is not the most fundamentally complex operation, but is the most extensively used operation (especially in signal processing). Innumerable schemes have been proposed for realization of the operation. Previously multiplication was implemented by repetitive sequence of addition, subtraction and shift operations [1] [2]. A system's performance is generally determined by the performance of the multiplier because the multiplier is generally the slowest element in the system [7].

Multiplication is a mathematical operation in which an integer is added a specified number of times. A number (multiplicand) is added to itself a number of times as specified by another number (multiplier) to form a result (product). In basic method of multiplication, firstly the multiplicand is placed on top of the multiplier. The multiplicand is then multiplied by each digit of the multiplier beginning with the rightmost, least significant digit (LSD). Intermediate results (partial-products) are placed one atop the other, offset by one digit to align digits of the same weight. The final product is determined by summation of all the partial-products. Basic multiplication technique is shown in figure:

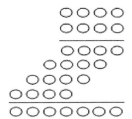

Fig.4.1.1.1 Basic Multiplication

In the binary number system the digits, called bits, are denoted by the set [0, 1]. The result of multiplying any binary number by a single binary bit is either 0 or the number itself. Hence forming the intermediate partial-products is simple and efficient. Summing these partial-products is the time consuming task for binary multipliers. The two main categories of binary multiplication include signed and unsigned numbers. The entire process of multiplication is divided in 3 parts:

1) Partial Product Generation.

2) Partial Product Reduction.

3) Final Adder.

4.1.2   Radix-2 Booth Multiplication Algorithm:

The Booth algorithm was invented by A. D. Booth forms the base of Signed number multiplication algorithms that are simple to implement at the hardware level, and that have the potential to speed up signed multiplication considerably. Booth's algorithm is based upon recoding the multiplier, y, to a recoded, value, z, leaving the multiplicand, x, unchanged. In Booth recoding, each digit of the multiplier can assume negative as well as positive and zero values.

Booth (Radix-2) algorithm provides a procedure for multiplying binary integers in signed-2's complement representation. For implementing Booth algorithm most important step is Booth recoding. The Booth recoding procedure is as follows:

1) Extend the sign bit 1 position if necessary to ensure that n is even.

2) Append a 0 to the right of the LSB of the multiplier.

3) According to the value of each recoded bit, each Partial Product will be 0, +M or –M.

The functional operation and recoding rules of Radix-2 algorithm are given in Table 1.

Table 1: Radix-2 recoding rules

Bits	Recoded bits	Partial Product
00	0	0
01	+1	1x Multiplicand
10	-1	-1x Multiplicand
11	0	0

4.1.3 Radix-4 Booth Multiplication Algorithm:

The original version of Booth algorithm (Radix 2) for multiplication was invented by, Andrew D. Booth. This algorithm had two drawbacks which are overcome by using modified Booth's (Radix 4) algorithm. This algorithm scans strings of three bits of multiplier at a time. There is no consistent definition of Radix in literature. In this thesis the radix is defined as 2k, where k is the number of bits, processed in iteration during the multiplication. In other words, the base of the number is changed to the radix 2k.

Generally, the multiplication algorithm is very similar to the operation made by hand. Thus, in a radix-4 algorithm, first we make a series of products between the multiplicand, and two bits of the multiplier generating in this way a set of words called partial products. There are several ways to increase multiplication speed. One is to reduce the number of partial products (PP) computed as part of the multiplication process. Another method is speeding up the addition of partial product array. Following example shows radix-4 multiplication in dot notation:

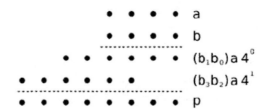

Fig. 4.1.3.1 Radix-4 Multiplication in Dot Notation

So for a radix-4 multiplication two bits a time are needed to form the partial product.

Radix-4 booth encoder performs the process of encoding the multiplicand based on multiplier bits. It will compare 3 bits at a time with overlapping technique. Grouping starts from the LSB, and the first block only uses two bits of the multiplier and assumes a zero for the third bit.

The number of bits multiplier/multiplicand is composed of gives exact number of partial products generated in multiplication operation. So, to perform the addition of partial product is main bottleneck in multiplication operation and considered as the important factor to speed up multiplication. In Booth's recoding (Radix-2) algorithm, if 2 'n' bit numbers are multiplied then 'n' partial products will be generated. The desired high speed can be achieved if the partial products are reduced. Modified Booth's (Radix 4) Algorithm uses the technique of partial product reduction to speed up multiplication operation. So, if 2 even 'n' bit numbers are multiplied, the number of partial products generated is 'n/2' and if 'n' is odd, number of partial products are 'n+1/2'.Thus, in Radix 4 the number of partial products is reduced to half. To have high speed multipliers, Modified Booth's Algorithm is an ultimate solution. This algorithm scans strings of three bits at a time.

The numbers of steps involved in Radix-4 multiplication algorithm are shown below:

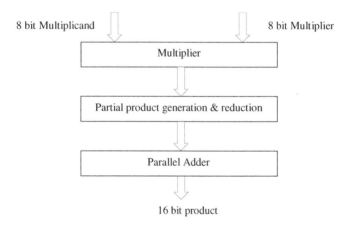

Fig. 4.1.3.2 Steps for Radix-4 multiplication

In Modified Booth's (Radix 4) Algorithm, the multiplicand is recoded based on bits of multiplier which can take any value from ± 1, ± 2 or 0. Three bits of multiplier are compared at a time, by using overlapping technique [4] [7]. Similar to Radix 2, we have to group bits of multiplier starting from LSB for which first block only uses two bits, considering third bit as zero.

The basic steps which are followed in this algorithm are given below:

1) Extend the sign bit 1 position if necessary to ensure that n is even.

2) Append a 0 to the right of the LSB of the multiplier.

3) According to the value of each recoded bit, each Partial Product will be 0, +M, -M, +2M or -2M [3].

The negative values of B are made by taking the 2's complement. The multiplication of M is done by shifting M by one bit to the left. Thus, in any case, in designing n bit parallel multiplier, only n/2 partial products are produced.

The functional operation and recoding rules of Radix-4 algorithm are given in Table 1.

Table 2: Radix-4 recoding rules

Bits	Recoded bits	Partial Product
000	0	0
001	1	1x Multiplicand
010	1	1x Multiplicand
011	2	2x Multiplicand
100	-2	-2x Multiplicand
101	-1	-1x Multiplicand
110	-1	-1x Multiplicand
111	0	0

4.1.4   Comparison of Radix-2 and Radix-4 Algorithms

- The shortcoming of Radix 2 Booth algorithm is that it becomes inefficient when there are isolated 1's. For example, 001010101(decimal 85) gets reduced to 01-11-11-11-1(decimal 85), requiring eight instead of four operations. This problem can be overcome by using high radix Booth algorithms.

- As we move towards Radix 8, less number of partial products are generated but more number of operations are required to generate {+1,+2, +3, +4, -1, -2, -3, -4}. In Radix 4 we need to save {+2, -2, +1, -1}

- With increase in radix-r, both the gate level and the delay performances will increase due to the complexity of the partial products encoded.
- The comparison of the 32-bit Radix-based Booth encoding variants indicates that the Radix-4 Booth Encoding multiplier is the best multiplier in terms of high-speed applications and low area constraint.

4.2 ADDITION AND SUBTRACTION:

This chapter deals with two another basic arithmetic operations based on higher radix (quaternary) number system, in which radix used is 4.

4.2.1 Basics of Addition and Subtraction:

Now-a-days adders are mostly used in various electronic applications such as Digital signal processors and computing devices these adders are used to perform various algorithms like FIR, IIR etc. In Modern electronics, Digital systems play a prominent role in day to day life. Arithmetic operations such as addition, subtraction and multiplication still suffer from known problems including limited number of bits, propagation time delay, and circuit complexity. The speed of digital processor depends heavily on the speed of adders they have constraints like area, power and speed requirements. The delay in an adder is dominated by the carry chain.

Modern computers are based on binary number system (radix =2). It has two logical states '0' and '1'. In such system, '1' plus '1' is '0' with carry '1' (i.e. 1+1=10). This carry should have to add with another '1', as a result further carry '1' generates. This creates the delay problem in computer circuits. In adders Binary Signed Digit Numbers are known to allow limited carry propagation with more complex addition process sand very large circuit for implementation. Some of the limitations of this system are computational speed which limits formation and propagation of carry especially as the number of bits increases. Therefore it provides large complexity and low storage density.

With the binary number system, the speed of arithmetic operations is limited by formation and propagation of the carry [11]. The basic operation of the multiplication is the addition. Therefore, fast adders are needed to build fast multipliers. There are many different fast adder designs. The most important ones are carry-propagation adder, carry-look-ahead adder, carry-skip adder, and carry-save adder. However, the architecture of fast adders always results in higher hardware complexity.

Most of the arithmetic operations suffer from problems like limited number of bits, circuit's complexity and delay [11]. Among the above mentioned adders carry look ahead adder produces less delay, but this adder is limited to small number of bits. So, the Radix-4 based adders are capable of performing carry free addition and borrow free subtraction and offers high speed.

The subtraction circuitry is nothing but the same circuit as the addition only with the difference of changing the sign bits of the second operand (as the subtraction can be performed by 2's complement addition of the second operand to the first).

In this thesis special higher radix (radix=4) based (quaternary) representation is used to perform carry free addition and borrow free subtraction. A special higher radix-based (quaternary) representation of binary signed-digit numbers not only allows carry-free addition and borrow-free subtraction but also offers other important advantages such as simplicity in logic and higher storage density. Carry free arithmetic operations can be achieved using a higher radix number system such as Quaternary Signed Digit (QSD). QSD number system eliminates carry propagation chain which reduces the computation time substantially, thus enhancing the speed of the machine. QSD Adder or QSD Multiplier circuits are logic circuits designed to perform high-speed arithmetic operations. A higher radix based signed digit number system, such as quaternary signed digit (QSD) number system, allows higher information storage density, less complexity. A high speed area effective adders and multipliers can be implemented using this technique. The advantage of carry free addition offered by QSD numbers is exploited in designing a fast adder circuit. Additionally adder designed with QSD number system has a regular layout which is suitable for VLSI implementation which is the great advantage.

4.2.2 Quaternary Signed Digit Number System:

Quaternary is the base 4 numeral system for which radix used is 4. In QSD number system carry propagation chain are eliminated which reduce the computation time substantially, thus enhancing the speed of the machine. As range of QSD number is from -3 to 3, the addition result of two QSD numbers varies from -6 to +6. The decimal numbers in the range of -3 to +3 are represented by one digit QSD number. As the decimal number exceeds from this range, more than one digit of QSD number is required. For the addition result, which is in the range of -6 to +6, two QSD digits are needed. In the two digits QSD result the LSB digit represents the sum bit and the MSB digit represents the carry bit. To prevent this carry bit to propagate from lower digit position to higher digit position QSD number representation is used. QSD numbers allow redundancy in the number representations. The same decimal number can be represented in more than one QSD representations. So such QSD represented number which prevents further rippling of carry is chosen. Signed digit representation of number indicates that digits can be prefixed with a – (minus) sign to indicate that they are negative. Signed digit representation can be used to accomplish fast addition of integers because it can eliminate carries.

Depending upon the radix number R, the number system are named as ternary (R = 3), quaternary (R = 4) etc. Ternary logic is based on ternary number system. Quaternary logic is based on Quaternary number system. Quaternary is the base 4 redundant number system. The degree of redundancy usually increases with the increase of the radix. The signed digit number system allows us to implement parallel arithmetic by using redundancy. QSD numbers are the Signed Digit numbers with the digit set as: {-3, -2, -1, 0, 1, 2, 3} respectively.

4.2.3   QSD number representation:

QSD numbers are represented using 3 bit 2's complement notation. Each number can be represented by

$$D = \sum_{i=0}^{n-1} x_i 4^i \qquad \qquad \text{... (1)}$$

Where x_i can be any value from the digit set {-3,-2,-1, 0, 1, 2, 3} for producing an appropriate decimal representation. A QSD negative number is the QSD complement of QSD positive number. For digital implementation, large number of digits such as 64, 128, or more can be implemented with constant delay. A high speed and area effective adders can be implemented using this technique. A higher radix based signed digit number system, such as quaternary signed digit (QSD) number system allows higher information storage density, less complexity, fewer system components and fewer cascaded gates and operations.

QSD number system allows multiple representations of any integer quantity. Example of 4 digit QSD number is shown below:

$(1332)_{QSD} = (1*4^3) + (3*4^2) + (3*4^1) + (2*4^0)$

$\qquad = 64+48+12+2$

$\qquad = 126$

The basic quaternary operators are very similar to binary operators the can be obtained from Boolean algebra.

4.2.4 Technique for Conversion from Binary to Quaternary:

1 digit QSD number can be represented by using a 3-bit binary equivalent are:

Table 3: Equivalent Binary Representation of QSD Numbers

QSD Number	Equivalent Binary Representation
-3	101
-2	110
-1	111
0	000
1	001
2	010
3	011

The 3q-bit binary data is converted from the n-bit binary data, thus equivalent q- digit QSD data is getting from conversion of n-bit binary data. We must split the $3^{rd},5^{th},7^{th}$ bit i.e. odd bit(from LSB to MSB)into two portion .we con not split the MSB bit this is major thing, for the odd bit is 0 then it is splited into 0&0 and if it is 1 then it is splited into 1&0.for example:$(1001110)_2$.

1 0 0 1 1 1 0

1 0 0 0 1 1 0 1 0

So we have to split the given binary data q-times. For example, the splitting is 1 time for conversion of a 2-digit quaternary number; the splitting is 2 times for conversion of a 3-digit quaternary number and so on. In each splitting one extra bit is generated. For conversion from binary to QSD, the required numbers of binary bits are,

$$n = 3q \{1*(n-1)\}$$

So, for converting binary number into its equivalent QSD the number of bits should be 3,5,7,9 etc. According to the above equation every 3-bit can be converted into its QSD.

Examples for conversion from decimal to quaternary number:

1) Let $(-179)_{10} = (101001101)_2$ have to be converted into its equivalent QSD. It has 9 binary bits. Its 3rd bit is 1,5^{th} bit is 0,7^{th} bit is 1.So from above equation, its equivalent QSD is of 4-digit.i.e,

1 0 1 0 0 1 1 0 1

1 0 1 0 0 0 0 1 1 0 0 1

Hence equivalent QSD of $(1\ 0\ 1\ 0\ 0\ 1\ 1\ 0\ 1)_2$ is $(-3\ 0\ 3\ 1)_{QSD}$.

4.2.5 Basic concept of QSD Addition and Subtraction:

In QSD number system carry propagation chain are eliminated which reduce the computation time substantially, thus enhancing the speed of the machine. As range of QSD number is from -3 to 3, thus addition result of two QSD numbers varies from -6 to +6. For performing any operation in QSD, first convert the binary number into quaternary signed digit. The basic idea of QSD (Radix=4) addition is depicted in following figure:

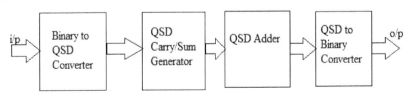

Fig 4.2.5.1: General Block Diagram for QSD Addition

4.2.6 Steps for QSD Addition:

To perform carry free addition, the addition of two QSD numbers can be done in two steps :

Step 1: First step generates an intermediate carry and intermediate sum from the input QSD digits i.e., addend and augend.

Step 2: Second step combines intermediate sum of current digit with the intermediate carry of the lower significant digit. So the addition of two QSD numbers is done in two stages. First stage of adder generates intermediate carry and intermediate sum from the input digits. Second stage of adder adds the intermediate sum of current digit with the intermediate carry of lower significant digit.

4.2.7 Rules for QSD Addition:

To remove the further rippling of carry there are two rules to perform QSD addition in two steps:

Rule 1: First rule states that the magnitude of the intermediate sum must be less than or equal to 2 i.e., it should be in the range of -2 to +2.

Rule 2: Second rule states that the magnitude of the intermediate carry must be less than or equal to 1 i.e., it should be in the range of -1 to +1.

By considering the Rules for carry free addition, the representation of QSD Number is shown in further table. According to these two rules the intermediate sum and intermediate carry from the first step QSD adder can have the range of -6 to +6. But by exploiting the redundancy feature of QSD numbers, such QSD represented number satisfies the above mentioned two rules.

When the second step QSD adder adds the intermediate sum of current digit, which is in the range of -2 to +2, with the intermediate carry of lower significant digit, which is in the range of -1 to +1, the addition result cannot be greater than 3, i.e., it will be in the range of -3 to +3. The addition result in this range can be represented by a single digit QSD number; hence no further carry is required. In the step 1 QSD adder, the range of output is from -6 to +6 which can be represented in the form of intermediate carry and intermediate sum in QSD format as shown in table 4.

It can be seen in the first column of Table 4 that some numbers have multiple representations, but only those that meet the above defined two rules are chosen. The chosen intermediate carry and intermediate sum are listed in the last. To prevent this carry bit to propagate from lower digit position to higher digit position QSD number representation is used. QSD numbers allow redundancy in the number representations. The same decimal number can be represented in more than one QSD representations. So such QSD represented number which prevents further rippling of carry is chosen. In the step 1 QSD adder, the range of output is from -6 to +6 which can be represented in the intermediate carry and sum in QSD format as shown in table 4.

Table 4: The outputs of all possible combinations of a pair of Addend (A) and Augend (B)

	-3	-2	-1	0	1	2	3
-3	-6	-5	-4	-3	-2	-1	0
-2	-5	-4	-3	-2	-1	0	1
-1	-4	-3	-2	-1	0	1	2
0	-3	-2	-1	0	1	2	3
1	-2	-1	0	1	2	3	4
2	-1	0	1	2	3	4	5
3	0	1	2	3	4	5	6

If the representation is restricted such that the intermediate carry is limited to a maximum of 1, and the intermediate sum is restricted to be less than 2, then the final addition will become carry free. Both inputs and outputs can be encoded in 3-bit 2's complement binary number. The mapping between the inputs, add end and augend, and the outputs, intermediate carry and sum are shown in binary format in Table 3.2.

Table 5: The Intermediate Carry and Sum between -6 to +6

Sum	Possible QSD Representation	QSD coded number
-6	$\bar{2}2, \bar{1}\bar{2}$	$\bar{1}\bar{2}$
-5	$\bar{2}3, \bar{1}\bar{1}$	$\bar{1}\bar{1}$
-4	$\bar{1}0$	$\bar{1}0$
-3	$\bar{1}1, 0\bar{3}$	$\bar{1}1$
-2	$\bar{1}2, 0\bar{2}$	$0\bar{2}$
-1	$\bar{1}3, 0\bar{1}$	$0\bar{1}$
0	00	00
1	$01, 1\bar{3}$	01
2	$02, 1\bar{2}$	02
3	$03, 1\bar{1}$	$1\bar{1}$
4	10	10
5	$11, 2\bar{3}$	11
6	$12, 2\bar{2}$	12

CHAPTER 5
FIELD PROGRAMMABLE GATE ARRAY

This chapter introduces about the FPGA concepts and FPGA Synthesis Flow. An FPGA is a device that consists of thousands or even millions of transistors connected to perform logic functions. They perform functions from simple addition and subtraction to complex digital filtering and error detection and correction.

5.1 Introduction to FPGA

A field programmable gate array (FPGA) is a semiconductor device that can be configured by the customer or the designer after manufacturing hence the name "field- programmable". Field Programmable gate arrays (FPGAs) are truly revolutionary devices that blend the benefits of both hardware and software. FPGAs are programmed using a logic circuit diagram or a source code in Hardware Description Language (HDL) to specify how the chip will work. They can be used to implement any logical function that an Application Specific Integrated Circuit (ASIC) could perform but the ability to update the functionality after shipping offers advantages for many applications. FPGAs contain programmable logic components called "logic blocks", and a hierarchy of reconfigurable interconnects that allow the blocks to be "wired together" somewhat like a one chip programmable breadboard. Logic blocks can be configured to perform complex combinational functions or merely simple logic gates like AND and XOR. In most FPGAs, the logic block also includes memory elements, which may be simple flip flops or more complete blocks of memory.

FPGAs blend the benefits of both hardware and software. They implement circuits just like hardware performing huge power, area and performance benefits over software, yet can be reprogrammed cheaply and easily to implement a wide range of tasks. Just like computer hardware, FPGAs implement computations spatially, simultaneously computing millions of operations in resources distributed across a silicon chip. Such systems can be hundreds of times faster than microprocessor-based designs. However unlike in ASICs, these computations are programmed into a chip, not permanently frozen by the manufacturing process. This means that an FPGA based system can be programmed and reprogrammed many times. FPGAs are being incorporated as central processing elements in many applications such as consumer electronics, automotive, image/video processing military/aerospace, base stations, networking/communications, super computing and wireless applications.

As the FPGA architecture evolves and its complexity increases. Today, most FPGA vendors provide a fairly complete set of design tools that allows automatic synthesis and compilation from design specifications in hardware specification languages, such as Verilog or VHDL, all the way down to a bit stream to program FPGA chips. Field Programmable Gate Arrays (FPGAs) are one of the fastest growing segments of the semiconductor industry [21]. They were first introduced in 1985, and since then they have quickly gained widespread acceptance as an excellent technology for implementing moderately large digital circuits in low production volumes. FPGAs are programmable devices that can be directly configured by the end user without the use of an integrated circuit fabrication facility. They offer the designer the benefits of custom hardware, eliminating high development costs and manufacturing time. Figure (4.3) shows a conceptual diagram of a typical FPGA. Field Programmable Gate Arrays are called this because rather than having a structure similar to a PAL or other programmable device, they are structured very much like a gate array ASIC (Application Specific Integrated Circuit).

The first programmable device was the programmable array logic (PAL). One of the PAL devices is PLD. Programmable Logic Devices (PLDs) are programmable devices that can be configured for a wide variety of applications. They enable faster implementation and emulation of circuit designs on hardware. The flexibility provided by these devices through the presence of reconfigurable elements has increased their popularity. There are two major types of PLDs: Field Programmable Gate Arrays (FPGAs) and Complex Programmable Logic Devices (CPLDs). Among the various possible FPGA architectures, lookup-table (LUT) based FPGA architectures have been the most popular ones. A LUT-based FPGA consists of an array of programmable logic blocks (PLBs) together with programmable interconnections. The maximum numbers of gates in an FPGA are as high as 500,000 and doubling every 18 months while the prices dropped dramatically.

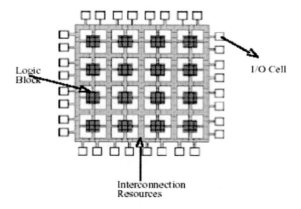

Figure 5.1.1: Conceptual diagram of a typical FPGA

Although there are many types of FPGAs, all architectures include logic blocks, I/O blocks, and programmable routing, which are arranged in a regular pattern. FPGA provide narrow logic resources; in other words, their logic blocks are generally small and uncommitted. One advantage of an FPGA over other types of FPDs is that they generally have much higher logic capacities than other FPDs and offer a higher ratio of flip flops to logic. A higher ratio of flip-flops to logic is important because flip-flops are often the limiting factor in designs. FPGAs are the most common form of FPD offered by programmable logic vendors. One such vendor, Xilinx, offers several different "families" of FPGAs that target different design sizes, design speeds, and cost requirements. Some of the more popular devices include the XC4000, the Spartan series, and the Virtex II series.

5.2 Basics about Advanced FPGA Trainer

The ADVANCED FPGA Trainer FPGA-A is useful to realize and verify various digital designs. User can construct VHDL/Verilog code and verify the results by implementing physically into the target device FPGA (SPARTAN XC3S-400-PQ208). With the help of this trainer user can simulate/observe various input and output conditions to verify the implemented design using the prototype board Advanced FPGA Trainer. Also user can select various I/O standard interfaces to the device.

5.2.1 Features:

- Useful to physically verify simple digital designs using FPGA
- A board that uses a SPARTAN-3 FPGA series device
- On board Function Generator
- Modular organization of circuit functions

5.2.2 PROGRAMMABLE LOGIC DEVICES [PLDs]

A Programmable Logic Device is a device whose logic characteristics can be changed and manipulated or stored through programming.

DIFFERENT TYPES OF PLDs:

➢ PROGRAMMABLE ARRAY LOGIC [PALs]

The most common and simple device that falls in this category is the PAL, which simply consists of an array of AND gates and an array of OR gates. The AND array is programmable while the OR array is relatively fixed.

➢ COMPLEX PROGRAMMABLE LOGIC DEVICES [CPLDs]

CPLD's are made up of smaller common Macro cells, which are programmable. CPLD's consists of multiple PAL like function block that can be interconnected through a switch matrix. These are **[Flash] EPROM based devices** i.e. these devices store their configuration even when power is switched off. Hence they need not to be configured every time when power is applied.

➢ FIELD PROGRAMMABLE GATE ARRAYS [FPGAs]

FPGA's are arrays of logic blocks, which can be linked together to form complex logic implementations. They are separated into two categories Fine Grained and Coarse Grained. Fine Grained being made up of sea of gates or transistors or small macro cells, while Coarse Grained being made up of bigger macro cells which are often made up of flip-flops and Look up Tables which make up the Combinational logic functions. These are RAM based devices i.e. these devices lose their configuration when power is switched off. Hence they have to be configured every time when power is applied.

➤ **APPLICATION SPECIFIC INTEGRATED CIRCUITS [ASICs]**

ASIC's are nothing but prefabricated pre-doped silicon chips. These are application specific designs. They cannot be reconfigured once manufactured. Once the design is completely finalized, it can be made as ASIC. Design changes are not possible but the size and speed is more.

5.3 Overview of the Xilinx Spartan-3 devices

Advanced FPGA Trainer uses Xilinx Spartan-3 family FPGA devices (XC3S-400). Based on the ratio between the number of logic cells and the I/O counts, the family is further divided into several subfamilies. Our discussion applies to all the subfamilies.

- **Logic cell, slice, and CLB**

The most basic element of the Spartan-3 device is a logic cell (LC), which contains a four-input LUT and a D FF. In addition, a logic cell contains a carry circuit, which is used to implement arithmetic functions, and a multiplexing circuit, which is used to implement wide multiplexers.

The LUT can also be configured as a 16-by-1 static random access memory (SRAM) or a 16-bit shift register. To increase flexibility and improve performance, eight logic cells are combined together with a special internal routing structure. In Xilinx terms, two logic cells are grouped to form a slice, and four slices are grouped to form a configurable logic block (CLB).

- **Macro cell**

The Spartan-3 device contains four types of macro blocks: combinational multiplier, block RAM, digital clock manager (DCM), and input/output block (IOB). The combinational multiplier accepts two 18-bit numbers as inputs and calculates the product. The block RAM is an 18K-bit synchronous SRAM that can be arranged in various types of configurations. A DCM uses a digital-delayed loop to reduce clock skew and to control the frequency and phase shift of a clock signal. An IOB controls the flow of data between the device's I/O pins and the internal logic. It can be configured to support a wide variety of I/O signaling standards.

- **Devices in the Spartan-3 subfamily**

Although Spartan-3 FPGA devices have similar types of logic cells and macro cells, their densities differ. Each subfamily contains an array of devices of various densities. The numbers of

LCs, block RAMS, multipliers, and DCMs of the devices from the Spartan-3 subfamily are summarized in table given below.

Table 6: Devices in Spartan-3 Family

Device	Number of LCs	Number of Block RAMs	Block RAM bits	Number of Multipliers	Number of DCMs
XC3S50	1728	4	72K	4	2
XC3S200	4320	12	216K	12	4
XC3S400	8064	16	288K	16	4
XC3S1000	17,280	24	432K	24	4
XC3S1500	29,952	32	576K	32	4
XC3S2000	46,080	40	720K	40	4
XC3S4000	62,208	96	1728K	96	4
XC3S5000	74,880	104	1872K	104	4

CHAPTER 6

RESULTS AND DISCUSSIONS

Following figures show the simulated output waveform for multiplication operation of signed as well as unsigned numbers. The multiplication results are obtained for two 8 bit numbers.

Fig.6.3 shows the multiplication result of two unsigned numbers, which is a positive number. Similarly, Fig.6.4 shows the multiplication result of signed numbers, which is a negative number.

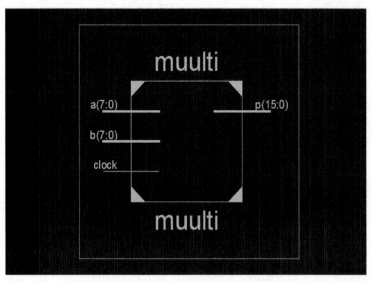

Fig. 6.1: RTL Schematic of Radix-4 multiplier

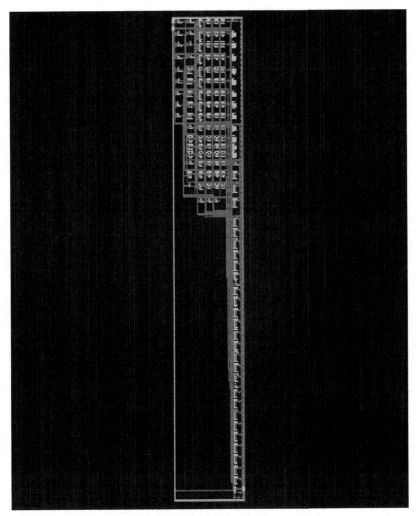

Fig.6.2: Internal RTL Schematic of Radix-4 multiplier

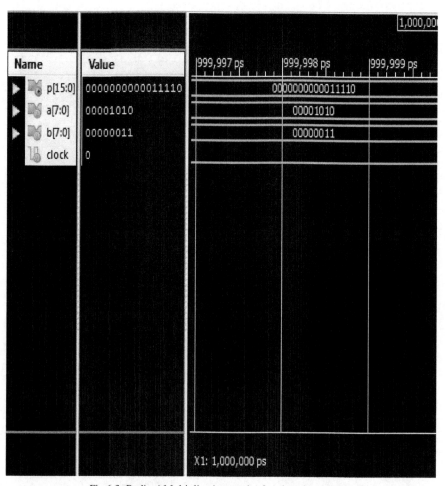

Fig.6.3: Radix-4 Multiplication result of unsigned number

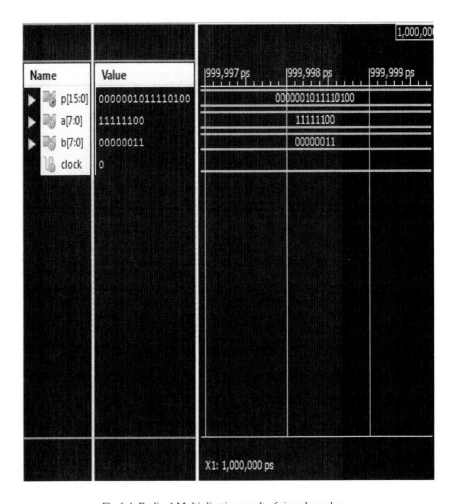

Fig.6.4: Radix-4 Multiplication result of signed number

Table 7: Device utilization of radix-4 booth multiplication

Logic Utilization	Used	Available	Utilization
Number of 4 input LUTs	169	7,168	2%
Number of occupied slices	86	3,584	2%
Number of slices containing only related logic	86	86	100%
Number of bonded IOBs	33	141	23%

The above table shows the device utilization summary for radix-4 booth multiplication operation. This shows the percentage of LUTs, slices & IOBs that has been utilized by multiplication operation.

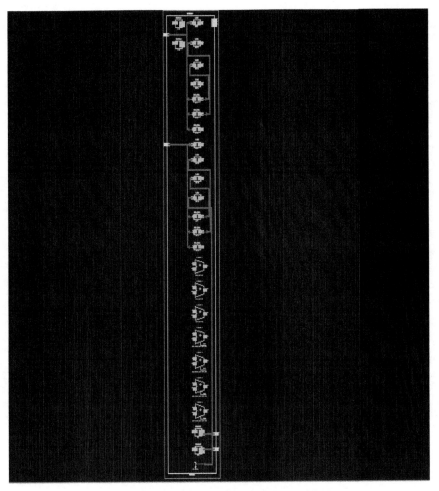

Fig. 6.5: Internal RTL schematic of Radix-4 addition

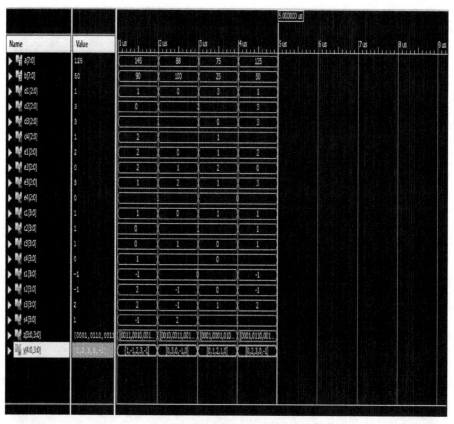

Fig. 6.6: Addition Result

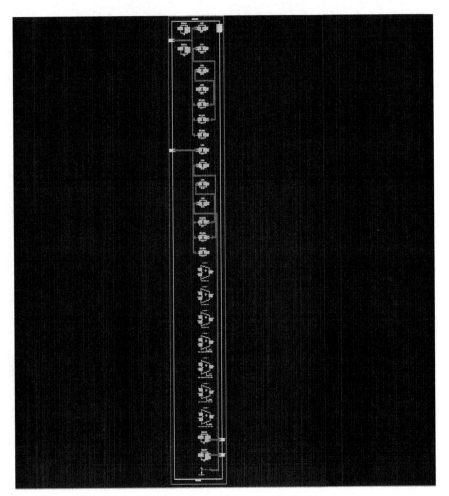

Fig. 6.7: Internal RTL schematic of Radix-4 subtraction

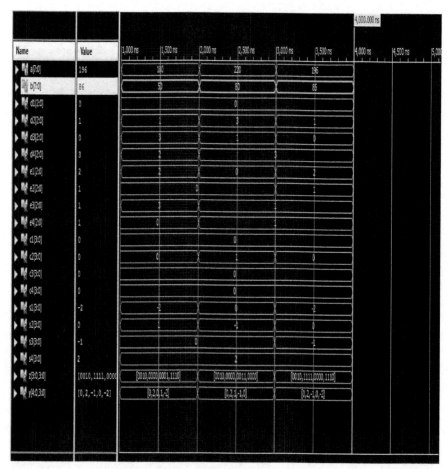

Fig.6.8: Subtraction Result

Table 8: Device utilization summary of addition/subtraction operations

Logic Utilization	Used	Available	Utilization
Number of slices	11	3584	0%
Number of 4 input LUTs	19	7168	0%
Number of bonded IOBs	36	141	25%

CHAPTER 7
CONCLUSION AND FUTURESCOPE

7.1 Conclusion

Radix-4 is the advanced and more sophisticated algorithm for designing the High Speed Multiplier for ALU using Minimal Partial Products. As compared to Radix 2 algorithm, Radix 4 algorithm halves the number of partial products and speed up the multiplication operation.

Addition and subtraction operations based on base-4 number system are both carry free and borrow free. Due to this feature, n-bit numbers can be easily added or subtracted without propagation of carry. Since there is no rippling of carry or borrow from previous stage to next stage, it helps to increase the speed of system.

7.2 Future Scope

Since the DSP processors are common in all digital electronic devices and the basic arithmetics plays vital role in such systems, these operations based on base-4 number system can be used to form co-processor. Such co-processor will help to compute above mentioned arithmetics and increase the speed of system.

BIBLIOGRAPHY

References:

1) Sukhmeet Kaur, Suman and Manpreet Signh Manna, "Implementation of Modified Booth Algorithm (Radix 4) and its Comparison with Booth Algorithm (Radix 2)", Advance in Electronic and Electric Engineering, Vol. 3, No.6, pp.683-690, 2013.

2) Shubhi Shrivastva, Pankaj Gulhane, "Optimized model of Radix-4 Booth Multiplier in VHDL", International Journal of Emerging Technology and Advanced Engineering, Vol.4, Issue 9, September 2014.

3) Prof .V .R. Raut, P. R .Loya, "FPGA Implementation of Low Power Booth Multiplier Using Radix-4 Algorithm", International Journal of Advanced Research in Electrical, Electronics and Instrumentation Engineering, vol.3, Issue 8, August 2014.

4) K. Babulu, G. Parasuram, "FPGA Realization of Radix-4 Booth Multiplication Algorithm for High Speed Arithmetic Logics", International Journal of Computer Science and Information Technologies, vol.2 (5), 2011.

5) Bodasingi Vijay Bhaskar, Valiveti Ravi Tejesvi, Reddi Surya Prakash Rao, "Implementation of Radix-4 Multiplier with a parallel MAC unit using MBE Algorithm", International Journal of Advanced Research in Computer Engineering and Technology, vol.1, Issue 5,July2012.

6) Wai-Leong Pang, Kah-Yoong Chan, Sew-Kin Wong, Choon-Siang Tan, "VHDL Modeling of Booth Radix-4 Floating Point Multiplier for VLSI Designer's Library" WSEAS TRANS. on SYSTEMS. Issue 12, Vol. 12, December 2013.

7) Rashmi Ranjan, Pramodini Mohanty, "A New VLSI Architecture of Parallel Multiplier Based on Radix-4 Modified Booth Algorithm using VHDL", International Journal of Computer Science and Engineering Technology (IJCSET), vol.3, No.4, April 2012.

8) Khalid Javeed, Xiaojun Wang, Mike Scott, "Serial and Parallel Interleaved Modular Multipliers on FPGA Platform", 2015 25th International conference on Field Programmable Logic and Applications (FPL), pp.1-4.

9) S.Shafiulla Basha, Syed. Jahangir Badashah, "Design and Implementation of Radix-4 Based High Speed Multiplier using Minimal Partial Products", International Journal of Advances in Engineering & Technology, vol.4, Issue 1, pp.314-325, July 2012.

10) A. B. Pawar, "Radix-2 Vs Radix-4 High Speed Multiplier", International Journal of Advanced Research in Computer Science and Software Engineering, Vol.5, Issue 3, March 2015.

11) Jyoti R. Hallikhed, Mahesh R.K., "VLSI Implementation of Fast Addition using Quaternary Signed Digit Number System", International Journal of Ethics in Engineering & Management Education, Vol.2, Issue 5, May 2015.

12) Shrikesh A. Dhakane, A.M.Shah, "FPGA Implementation of Fast Arithmetic Unit Based on QSD", International Journal of Computer Science and Information Technologies (IJCSIT), Vol. 3(5), 2014.

13) Prashant Y. Shende, Dr. R.V.Kshirsagar, "Quaternary Adder Design using VHDL", International Journal of Engineering Research and Applications (IJERA), Vol.3, Issue 3, May-June 2013, pp.270-273.

14) Snehal B. Sahastrabudhey, K.M. Bogawar, "Arithmetical Operations in Quaternary System Using VHDL", IJCSET, Vol 2, Issue 4, pp. 1160-1163, April 2012.

15) S. Jakeer Hussain, K. Sreenivasa Rao, "Design and implementation of Fast Addition Using QSD for Signed and Unsigned Numbers", International Journal of Engineering Research, Volume No. 3, Issue No: Special2, pp: 52-54, March 2014.

16) M Naveen Krishna, T Ravisekhar, "Fast Arithmetic operations with QSD using Verilog HDL", International Journal of Engineering Science and Innovative Technology (IJESIT), Volume 3, Issue 4, July 2014.

17) C. V. Sathish Kumar, P. Jaya Rami Reddy, "Implementation of Fast Adder Using QSD for Signed and Unsigned Numbers", International Journal of Science, Engineering and Technology Research (IJSETR), Volume 3, Issue 11, November 2014.

18) Kothuru. Ram Kumar, B. Praveen Kumar, "Fast Addition Using QSD VLSI Adder for Better Performance", International Journal of Trend in Research and Development, Volume 2(6), Nov-Dec 2015.

19) G. Manasa, M. Damodhar Rao, K. Miranji, "Design and Analysis of Fast Addition Mechanism for Integers using Quaternary Signed Digit Number System", International Journal of VLSI and Embedded Systems-IJVES, Vol 05, Article 09455, October 2014.

20) Srinivasasamanoj. R, M. Sri Hari, B. Ratna Raju, "High Speed VLSI Implementation of 256-bit Parallel Prefix Adders", International Journal of Wireless Communications and Networking Technologies, Volume 1, No.1, August-September 2012.

21) J. Bhasker, A Verilog HDL Primer, B.S.Publications, Third edition.

22) Doughas A. Pucknell, Kamran Eshraghian, Basic VLSI Design, PHI Learning Private Limited, Third Edition.

23) P.Pal Chaudhuri, Computer Organization and Design, PHI Learning Private Limited, Third Edition.

24) Wikipedia The Free Encyclopedia, http://www.wikipedia.org.in/edu.

Appendix

A. Radix-4 Multiplication:

Radix-4 multiplication is implemented in XILINX by using Booth algorithm. This algorithm generates (N/2) partial products and helps to increase the computation speed.

Radix-4 Multiplication Example:

Consider example for radix-4 Multiplication:

Multiplicand 0 0 0 0 1 0 1 0 $(10)_{10}$

Multiplier   0 0 0 0 0 0 1 1 0 $(3)_{10}$

In this the multiplier bit has to be recoded, this is as shown further:

Multiplicand 0 0 0 0 1 0 1 0

  Multiplier   0 0 0 0 0 0 1 1 0

      ↓ ↓ ↓ ↓

          0   0   1 -1

        0000000011110110

        00000000001010

        000000000000

        0000000000

Product 0000000000011110 $(30)_{10}$

B. Radix-4 Addition and Subtraction:

Similar to Radix-4 multiplication, Addition and Subtraction Operations are also implemented in XILINX by using QSDN system. Since these operations are free from carry and borrow, it requires less computation time and help to increase the computation speed.

Addition Example:

To perform QSD addition of two numbers A=143, B=100

First both numbers are represented in equivalent QSD numbers.

$(143)_{10} = (2\ 0\ 3\ 3)_4$ and $(100)_{10} = (1\ 2\ 1\ 0)_4$

A= $(143)_{10}$ 2 0 3 3

B= $(100)_{10}$ 1 2 1 0

_____

Sum=   3 2 4 3

_____

IC   1 0 1 1

IS $\bar{1}$ 2 0 $\bar{1}$

_____

O/p   1 $\bar{1}$ 3 1 $\bar{1}$

The sum output is $(1\ \bar{1}\ 3\ 1\ \bar{1})_4$ which is equal to $(243)_{10}$

Subtraction Example:

To perform QSD subtraction of two numbers A=143, B= -100

First both numbers are represented in equivalent QSD numbers.

$(143)_{10} = (2\ 0\ 3\ 3)_4$ and $(-100)_{10} = (\bar{1}\ \bar{2}\ \bar{1}\ 0)_4$

A= $(143)_{10}$ 2 0 3 3

B= $(-100)_{10}$ $\bar{1}$ $\bar{2}$ $\bar{1}$ 0

_____

Sum=   1 $\bar{2}$ 2 3

_____

IC   0 0 0 1

IS $1\ \bar{2}\ 2\ \bar{1}$

_____

O/p $0\ 1\ \bar{2}\ 3\ \bar{1}$

The sum output is $(0\ 1\ \bar{2}\ 3\ \bar{1})_4$ which is equal to $(43)_{10}$.

TOC

Printed in Great Britain
by Amazon

83044339R00031